The Library of Subatomic Particles™

The Proton

Fred Bortz

The Rosen Publishing Group, Inc., New York

To Susan, with every positive wish

Published in 2004 by The Rosen Publishing Group, Inc.
29 East 21st Street, New York, NY 10010

Library of Congress Cataloging-in-Publication Data

Bortz, Alfred B.
The proton / by Fred Bortz. — 1st ed.
    p. cm. — (The library of subatomic particles)
Summary: A look into the discovery of the most fundamental subatomic particle in nature, the proton, which determines why elements have different physical and chemical properties.
Includes bibliographical references and index.
ISBN: 978-1-4358-3664-8
1. Protons—Juvenile literature. 2. Particles (Nuclear physics)—Juvenile literature. [1. Protons. 2. Particles (Nuclear physics)]
I. Title. II. Series.
QC793.5.P72B67 2004
539.7'2123—dc22

                                                    2003013107

Manufactured in the United States of America

On the cover: An artist's illustration of a heavy atomic nucleus consisting of protons and neutrons.

Contents

Introduction

All matter is made of atoms and combinations of atoms called molecules. That statement may not seem surprising to you, but as recently as the beginning of the last century, serious scientists were still looking for proof that atoms and molecules existed. Atomic theory explained a lot about chemistry, but it was not until 1905 that Albert Einstein pointed out that the jiggling movement of dust particles in fluids, known as Brownian motion, was exactly what would be expected if tiny gas molecules were constantly bouncing off them.

The scientific theory of atoms goes back another hundred years to 1803, when John Dalton adopted an idea of the ancient Greek philosophers Leucippus and Democritus. Twenty-three centuries earlier, they imagined cutting up a piece of matter until it was *atomos*, meaning "indivisible," but that was long before modern science. Dalton was the first to bring the notion of atoms into the interpretation of laboratory observations.

In Dalton's theory, the atoms of Leucippus and Democritus turn out to be molecules. A water molecule is made of two hydrogen atoms and one oxygen atom and is therefore not indivisible. But it is the smallest unit of matter that can still be called water. Dalton spoke of elements, which are made of only one kind of atom, and compounds, which are made of only one kind of molecule.

Dalton's theory still considered atoms indivisible, but as more scientists discovered more elements, it was natural to wonder if the atoms were made of even smaller particles that distinguished one kind of atom from another. By the end of the nineteenth century, physicists (scientists who study matter and energy) began to answer those questions. Even before Einstein persuaded his fellow scientists that atoms were real, researchers had begun to find evidence of subatomic particles. This series tells the story of their discoveries, each book turning the spotlight on a different particle.

This book is the story of the proton, a particle that is found deep within every atom and is responsible for the energy that lights up the stars.

Chapter One

The Atomic Nature of Matter

Twenty-five centuries ago, when Leucippus and his student Democritus came up with the idea that matter was made up of tiny indivisible pieces, they never imagined where it would lead. Today's world of human-made substances depends on our knowledge of atoms, the subatomic particles within them, and the forces that hold them together or break them apart.

When the ancient Greeks turned their philosophical minds to the world around them, they began with two simple questions: What is matter made of, and why do different kinds of matter behave so differently from each other? When they imagined dividing everyday substances into smaller and smaller bits until the pieces were indivisible atoms, they also decided that the atoms of each substance would have a particular shape and texture. For instance, they concluded that water atoms would be round and smooth, while rock atoms would be hard and sharp or gritty.

The presumed atoms were much too tiny for the Greeks to test their ideas. Besides, the notion of testing theories by observation, a cornerstone of modern science, was not yet part of human culture. Toward the end of Democritus's life, Greek philosophy entered what is often called the golden era, where great thinkers such as Socrates, Aristotle, and Plato used their powerful minds and logic to deduce what they believed to be the truth about the natural world.

Philosophizing About Matter. The modern idea of atoms—that they are the smallest possible piece of a substance—goes back more than 2,500 years to the ancient Greek philosophers Democritus, shown here, and Leucippus.

Aristotle was so brilliant in many fields that his ideas were rarely questioned. For nearly 2,000 years, most people accepted his conclusion that all the world's matter was made of four elements: earth, air, fire, and water. The idea of atoms all but disappeared. Today we know that both Democritus and Aristotle were right in general but wrong in detail.

Democritus said that there was a limit to how small a piece of matter can be cut and still remain the same substance, and that was right. But most substances are compounds. The smallest possible piece of most substances is usually a molecule instead of an atom. Furthermore, molecules can be divided into their atoms, and even atoms are divisible. Aristotle was correct that all matter is made of combinations of particular elements, but not the four he wrote about. Water is a compound; earth and air are mixtures containing both elements and compounds; fire is a process that produces energy as atoms rearrange themselves into different compounds; and the number of natural elements is nearly a hundred, not four.

As you might imagine, the road from the philosophy of the ancient Greeks to today's scientific knowledge of atoms is a long one with many interesting twists and stops along the way. It begins with an activity called alchemy, in which people tried to make certain substances out of others, often by heating things and melting them together. Most often, alchemists were searching for ways to turn less valuable metals into gold using techniques that we now know

Flash and Dazzle. Though the modern science of chemistry has roots in the techniques of the ancient practice of alchemy, some alchemists were frauds. They dazzled audiences with spectacular displays that produced convincing illusions such as this one, where the alchemist claimed to be creating a living being from nonliving ingredients.

were doomed to fail. Gold is an element, and neither ancient alchemy nor modern chemistry can change one kind of atom into another.

Though many alchemists were fraud artists, others succeeded in developing a rudimentary knowledge of matter, extracting or purifying many useful elements and compounds from natural substances. By the seventeenth century, scientific thinking had begun to take hold, and

I've Got It! In this picture, John Dalton, whose 1810 book described the atomic nature of matter that is still the cornerstone of modern chemistry, seems to be pondering his great discovery.

alchemy was gradually transformed into the science of chemistry. Eighteenth-century chemists made a number of important discoveries, including facts about the behavior of gases, the processes of combustion and corrosion, and the relationship between electricity and matter. None of those phenomena were fully understood, but plenty of evidence and measurements were being gathered systematically and scientifically.

As the nineteenth century dawned, English scientist John Dalton (1766–1844), already in his thirties, was turning his attention from meteorology to chemistry, hoping to gain a better understanding of the gases of the air. He soon realized that the ancient idea of atoms could explain many of the phenomena that others had observed in gases and chemical reactions. In 1810, Dalton published a book that revolutionized chemistry. The book, *A New System of Chemical Philosophy*, was based on the assumption that all matter was made of atoms. Dalton explained that each element was made of a particular kind of atom, and all of its atoms were identical to each other. The book also stated that atoms of different elements have different properties, including their weight. When atoms join to form compounds, it is always in small numbers of whole atoms—no fractional atoms are allowed.

Using those simple rules, Dalton was able to determine the atomic weight of different elements. He assigned hydrogen, the lightest element, one unit of atomic weight and determined the atomic weight of other atoms from that. For instance, he knew that water was

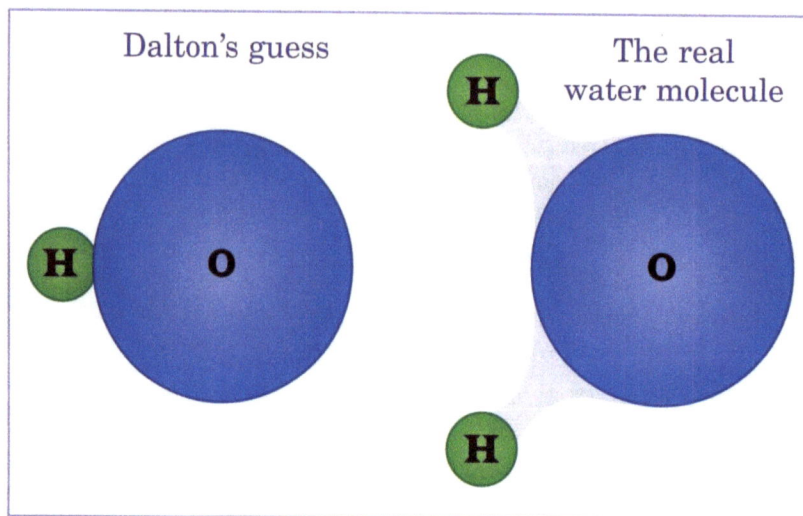

Dalton's guess

The real water molecule

H

H

O

O

H

Water You Doing? To determine the atomic weights of different substances, Dalton measured the way substances combined. His first guess for the atomic weight of oxygen was incorrect because he assumed that a water molecule was made up of one atom each of hydrogen and oxygen, rather than two hydrogen atoms and one oxygen atom.

a compound of hydrogen and oxygen, with eight times as much oxygen by weight. Assuming that a water molecule had one atom of each element, he set the atomic weight of oxygen at eight units. Later on, when more research showed that water molecules had two atoms of hydrogen and one of oxygen, scientists corrected that result, setting the atomic weight of oxygen at sixteen.

Following Dalton's method, scientists studying chemical reactions gradually identified more compounds and the elements that composed them, and they determined the atomic weights of each element. Though no one had detected individual atoms, Dalton's idea of elements and compounds, atoms and molecules, had given chemistry a new basic vocabulary. As is often the case with scientific breakthroughs, atomic theory also opened up a wealth of new questions, including these: How many elements are there, and is there a way to classify them to better understand their chemical behavior?

Periodic Properties

By 1869, a total of sixty-three elements were known, and scientists were having a hard time keeping track of them. They could see hints of similarities and patterns among the elements, such as in their melting or boiling points, their densities (how much each cubic centimeter weighs), the way they combined with other elements, and the properties of the compounds they formed. Still, no one had come up with a successful way to classify them—until, according to a famous story, Russian chemist Dmitry Ivanovich Mendeleyev (1834–1907) had a dream.

Periodic Table of the Elements

Representative (Main Group) Elements

Transition Metals

1 H 1.0079												
3 Li 6.941	4 Be 9.012											5 B 10.81
11 Na 22.990	12 Mg 24.305											13 Al 26.98
19 K 39.098	20 Ca 40.078	21 Sc 44.956	22 Ti 47.88	23 V 50.942	24 Cr 51.996	25 Mn 54.938	26 Fe 55.845	27 Co 58.933	28 Ni 58.69	29 Cu 63.546	30 Zn 65.39	31 Ga 69.72
37 Rb 85.468	38 Sr 87.62	39 Y 88.906	40 Zr 91.224	41 Nb 92.906	42 Mo 95.94	43 Tc 98	44 Ru 101.07	45 Rh 102.906	46 Pd 106.42	47 Ag 107.868	48 Cd 112.411	49 In 114.8
55 Cs 132.905	56 Ba 137.327	57 La 138.906	72 Hf 178.49	73 Ta 180.948	74 W 183.84	75 Re 186.207	76 Os 190.23	77 Ir 192.22	78 Pt 195.08	79 Au 196.967	80 Hg 200.59	81 Tl 204.38
87 Fr 223	88 Ra 226.025	89 Ac 227.028	104 Rf 261	105 Db 262	106 Sg 263	107 Bh 262	108 Hs 265	109 Mt 266	110 Uun 269	111 Uuu 272	112 Uub 277	

Rare Earth Elements

Lanthanides	58 Ce 140.115	59 Pr 140.908	60 Nd 144.24	61 Pm 145	62 Sm 150.36	63 Eu 151.964	64 Gd 157.25	65 Tb 158.925	66 Dy 162.5	67 Ho 164.93
Actinides	90 Th 232.038	91 Pa 231.036	92 U 238.029	93 Np 237.048	94 Pu 244	95 Am 243	96 Cm 247	97 Bk 247	98 Cf 251	99 Es 252

The Periodic Table. In 1869, Dmitry Mendeleyev was the first scientist to arrange the chemical elements in rows and columns that successfully displayed their similar properties. Because the properties followed a sequence that repeated itself, he called it the periodic table of the elements, the name we still use for today's arrangement, shown here. The original table had more spaces than the sixty-three elements then known. Mendeleyev correctly predicted not only that the gaps would be filled by undiscovered elements, but also what the properties of those elements would be from their placement in the table.

Representative (Main Group) Elements				1 H 1.0079
6 C 12.011	7 N 14.007	8 O 15.999	9 F 18.998	10 Ne 20.180
14 Si 28.086	15 P 30.974	16 S 32.066	17 Cl 35.453	18 Ar 39.948
32 Ge 72.61	33 As 74.922	34 Se 78.96	35 Br 79.904	36 Kr 83.8
50 Sn 118.71	51 Sb 121.76	52 Te 127.60	53 I 126.905	54 Xe 131.29
82 Pb 207.2	83 Bi 208.980	84 Po 209	85 At 210	86 Rn 222
114		116		118

68 Er 167.26	69 Tm 168.934	70 Yb 173.04	71 Lu 174.967
100 Fm 257	101 Md 258	102 No 259	103 Lr 262

Mendeleyev was a professor at St. Petersburg University known for his thorough knowledge of the elements and their properties. He also had a small country estate, and his neighbors relied on his advice about farming and cheese making. So from time to time, he would schedule a visit to the country. Traveling by train, he would often occupy his mind by playing the game of patience with a deck of cards. Like most solitaire games, the object of patience is to build an arrangement of the cards from the highest to the lowest within each suit (spades, hearts, diamonds, and clubs).

Before one such trip in early 1869, he had been

working almost day and night for three days, trying to discover a way to classify the elements that would make sense of all the fragments of patterns he and others had noticed. He decided to make a set of cards, one for each element, listing the known properties of each, and he arranged them in order of increasing atomic weight. Just before he was to leave to catch his train, the weary professor fell asleep and dreamed of playing patience with his deck of element cards. When he woke up, he knew what he had to do. The atomic weights were like the order of the cards. All he needed was to figure out how many groupings there were (nature had no reason to choose four suits as in playing cards) and how many cards were in each.

By the time Mendeleyev arrived at his destination, the arrangement had begun to fall into place. Though he had used other properties to develop his arrangement, he discovered that the groupings seemed to follow a chemical property called valence, which accounted for how many of one atom would combine with how many of another. For example, the elements called the alkali metals—lithium, sodium, potassium, rubidium, and cesium—all fell into alignment, as did the elements called halogens, which are fluorine, chlorine, bromine, and iodine. As atomic weight would increase, the atoms would follow a pattern: going from one valence to the next to the next and periodically starting over again. Mendeleyev thus called his arrangement the periodic table of the elements.

The table was not perfect, and it had a few gaps. Mendeleyev claimed that the gaps represented elements not yet discovered. He predicted not only that they would be found, but also what their atomic weights and density would be—and he was right! With those discoveries, the periodic table of the elements was established as one of the great ideas of chemistry. Like Dalton's atomic theory, Mendeleyev's periodic table led to important new questions, such as what makes the properties of atoms periodic? The discovery of subatomic particles was required before that question could be answered.

Chapter Two

Inside the Atom

The periodic table brought order to atomic theory, especially as more and more elements were discovered, but it still did not persuade all scientists that atoms actually existed. It's one thing to say that matter behaves as if it is made of atoms, and quite another to say that it is actually made of atoms. Then along came J. J. Thomson (1856–1940) of Cambridge University in England, who, in 1897, described his discovery of a tiny bit of matter that we now call the electron. It was less than a thousandth as heavy as the hydrogen atom, yet it had as much negative electrical charge as that atom might carry in its positive charge.

By then, scientists understood electricity quite well. They knew that two positively charged bodies or two negatively charged bodies would repel (push away from) each other, while a pair of bodies with opposite electric charges would attract each other. The attractive or repulsive forces get much stronger as the

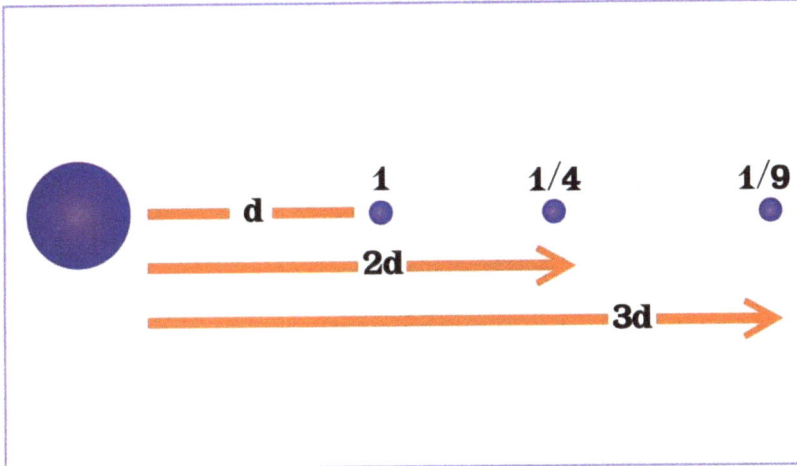

Electric Forces. The force between two electrical charges decreases as the separation between them increases. If they experience one unit of force when separated by a certain distance, then at twice that distance they would experience one-fourth of the force (a half of a half as much), and at three times that distance the force would be only one-ninth (a third of a third) as great.

charged bodies get closer together. Decreasing their separation to half its value multiplies the force by four (2×2). If the distance is shortened to one-third of its original value, the force increases by nine (3×3). They also knew that electricity was related to chemical reactions and valences and was thus probably important in atoms. Thomson's result suggested that atoms are not indivisible but are made of even smaller subatomic particles—tiny negatively charged electrons and much heavier positively charged particles—held together by electric forces.

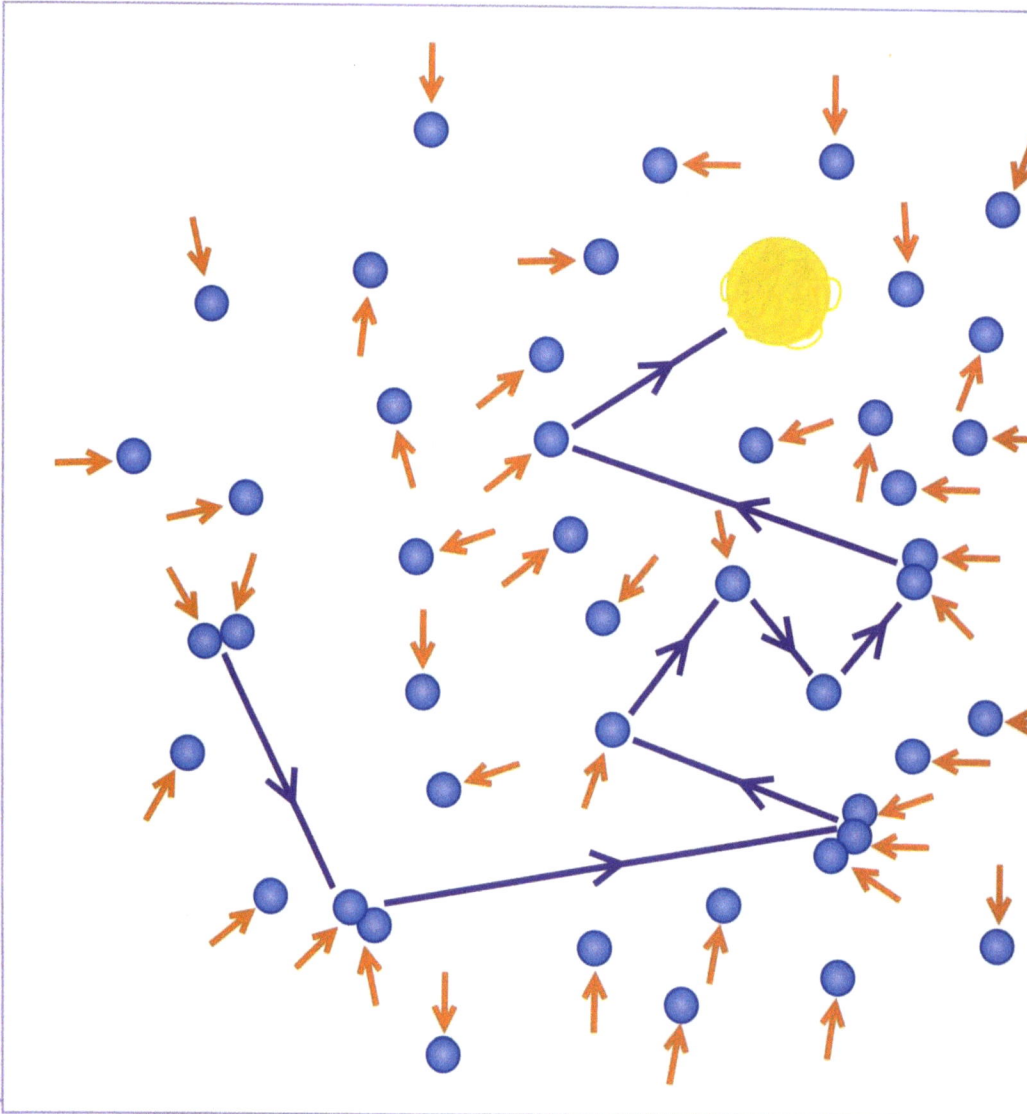

Brownian Motion. In 1827, botanist Robert Brown observed under a microscope a phenomenon that was soon named after him. He noticed that pollen grains suspended in water followed random jiggling paths like the ones shown here. In 1905, Albert Einstein's mathematical analysis showed that Brownian motion was exactly what would result from collisions between the grains and the atoms or molecules of the fluid. That was the first result that demonstrated the existence of individual molecules.

Still, many scientists felt that the atomic theory was incomplete without evidence of actual atoms. The problem was that they weren't looking in the right places. The evidence that atoms and molecules were real was already known but was not recognized as being due to atomic effects. It was called Brownian motion after the English botanist Robert Brown who observed in 1827 that pollen grains suspended in water under a microscope followed random jiggling paths, even if the water was absolutely still. Over the years, some scientists proposed that these grains were bumped about by molecules, and others studied Brownian motion more carefully, measuring the paths for different sizes of particles at different temperatures. Finally in 1905, Albert Einstein (1879–1955) calculated the motion that would be expected if atoms at a certain temperature collided with dust or pollen particles of a certain size, and the results matched Brownian motion perfectly. People couldn't see individual atoms, but they could see their combined effects.

From Plum Pudding to Planetary Models

So if atoms were real and contained tiny negatively charged electrons, what about the positively charged parts? Scientists knew that the heavier atoms were, the more electrons they had. The elements were soon labeled not only by their atomic weight but also by their atomic number—the number of electrons.

Since atoms are electrically neutral overall, they must contain equal positive and negative electric charges. Scientists wondered if the positively charged matter was in the form of individual particles or as a spread-out mass. J. J. Thomson put forward an educated guess. Since electrons carry so little mass, he envisioned the positively charged bulk of atoms as a kind of pudding containing tiny electron plums.

Thomson's plum pudding model was put to the test by Ernest Rutherford (1871–1937), who had come up with a way to probe matter by bombarding it with the emissions from radioactive substances. Rutherford had explored the recently discovered phenomenon of radioactivity when he was Thomson's student between 1895 and 1898. He found that radioactivity had two distinct forms, which he named alpha rays and beta rays. He then went on to become a professor at McGill University in Montreal, Canada. There, in 1902, he and his colleague Frederick Soddy discovered gamma rays, a third form of radioactivity. They also discovered that alpha and beta radiation were streams of fast-moving particles of opposite electric charge. The alphas were positively charged and much more massive than the negatively charged betas. (We now know that beta rays are electrons.)

Rutherford returned to England in 1907 as a professor at the University of Manchester, and he began shooting beams of alpha particles through metal foil and measuring how the alphas deflected, or scattered, as they interacted with atoms of the metal. By studying alpha scattering carefully, he hoped to be able to determine the size, spacing, and perhaps even the shape of the atoms in the foil. His student Hans Geiger devised an instrument to detect and count the alpha particles. They determined that alpha particles were helium atoms without their electrons, just as Rutherford had suspected.

In 1909, they began their scattering experiments, and the results were surprising. Nearly all the alphas passed straight through the foil or deflected only slightly. If the atoms were hard balls, Rutherford and Geiger would have expected more deflection. Also puzzling was this: A few alpha particles were unaccounted for. Geiger's counters had been placed behind the metal foil. Had some particles been deflected so much that they had missed the detectors? If so, what was scattering those few alpha particles at such large angles? While Geiger continued his detailed measurements, Rutherford assigned the task of looking for large-angle scattering to Ernest Marsden, another of his students. Marsden found that the missing alpha particles scattered to the left or right of the original detectors and a few even scattered backward!

Ernest Rutherford. In 1911, Rutherford announced the discovery of a planetary model of the atom based on experiments in which he bombarded thin metal foil with radioactive beams. He described atoms as consisting of heavy compact nuclei orbited by lightweight electrons under the influence of electric forces, just as gravity holds the planets in orbit around the sun. He later correctly predicted that the nucleus consisted of protons and neutrons.

Rutherford described this result as "almost as incredible as if you had fired a 15-inch [38-centimeter] shell at a piece of tissue-paper and it came back and hit you."

Rutherford explained his results with a new model of the atom. He described an atom as a miniature solar system with electrical forces playing the role of gravity. The atom is mostly empty space. Most of its mass is concentrated in a very small, positively charged nucleus (plural: nuclei) about a ten thousandth of the size of the atom! In orbit around that minuscule but massive "sun" are much tinier negatively charged "planets," the electrons. Because the atoms are mostly empty space, most alpha particles would pass through the foil without coming close enough to a nucleus to be scattered very much. Only on rare occasions would a fast-moving alpha particle make a nearly direct hit on a much heavier nucleus, which then scattered the alpha sideways or even backward.

Weighty Problems with the Periodic Table

But what was in the nucleus? Rutherford and other scientists concluded that the nucleus was composed of positive particles, which they called protons, each nucleus containing as many protons as that atom had electrons. Because it was not difficult to experimentally remove electrons from atoms, scientists decided that it was better to define the atomic number by the positive charge in the nucleus, that is, by the number of protons, instead of by the number of electrons. Hydrogen, the simplest atom with atomic number 1, has a nucleus with a single proton. Helium, with atomic number 2, has two protons, and so forth.

However, the issue proved to be more complicated. The atomic mass of hydrogen is one, but the atomic mass of helium is four. Other atoms reveal similar discrepancies. Lead, for example, has an atomic number of 82 and an atomic mass of approximately 207. Protons alone could not account for even half the mass of most nuclei. What else might be there?

Perhaps this extra mass could explain another problem in Rutherford's atomic model. The electric force, as we have seen, increases as the distance between charged bodies decreases. The protons, packed so tightly together in the nucleus, would feel enormous repulsive forces. Such powerful forces would blow the nucleus apart. Whatever else was in the nucleus, it not only gave the nucleus more mass but also exerted some force that held the nucleus together against the repulsive force of the protons. That meant that nature had powerful forces deep within the atom that physicists had not yet imagined.

Chapter Three

Inside the Nucleus

Even before Rutherford discovered the nucleus, his studies of radioactivity had made him very famous. His careful measurements of alpha, beta, and gamma rays led to the realization that nature was succeeding where alchemy had failed in transforming one element into another. Rutherford, his former colleague Frederick Soddy, and other scientists learned to identify different radioactive elements by distinct characteristics of their alpha, beta, and gamma rays.

Rutherford and Soddy were particularly interested in tracking the elements from their original form to their new forms. For instance, an atom that emits an alpha particle loses four units of atomic mass while its atomic number decreases by two. The atomic mass of another atom that emits a beta particle does not change, but its atomic number increases by one unit. In both cases, an atom of one element transforms, or "transmutes," into an atom of a different element. The new "daughter" atom is often radioactive, frequently more so than its "parent," so there is a chain of radioactive decay from one atom to another.

Rutherford and Soddy discovered that several of the daughter atoms in different radioactive decay chains behaved the same way chemically, which meant that they had the same atomic numbers, yet their atomic masses were different. Scientists called such atoms isotopes, and they soon discovered that many nonradioactive atoms also had more than one isotopic form. To distinguish one isotope from another, they began denoting atoms by their chemical symbol and their atomic mass. Chlorine, for example, has isotopes with atomic masses of 35 and 37. Cl^{35} is the more common of the two, which explains why chemists measure the atomic mass of natural chlorine as 35.47.

Having not yet discovered the nucleus, Rutherford did not realize that the radioactivity was coming from deep inside the atom or that transmutation was very different from alchemy and its scientific successor, chemistry. Chemical changes are related to the atoms' electrons and the bonds they form between atoms, but chemistry never changes atomic number or

Nucleus

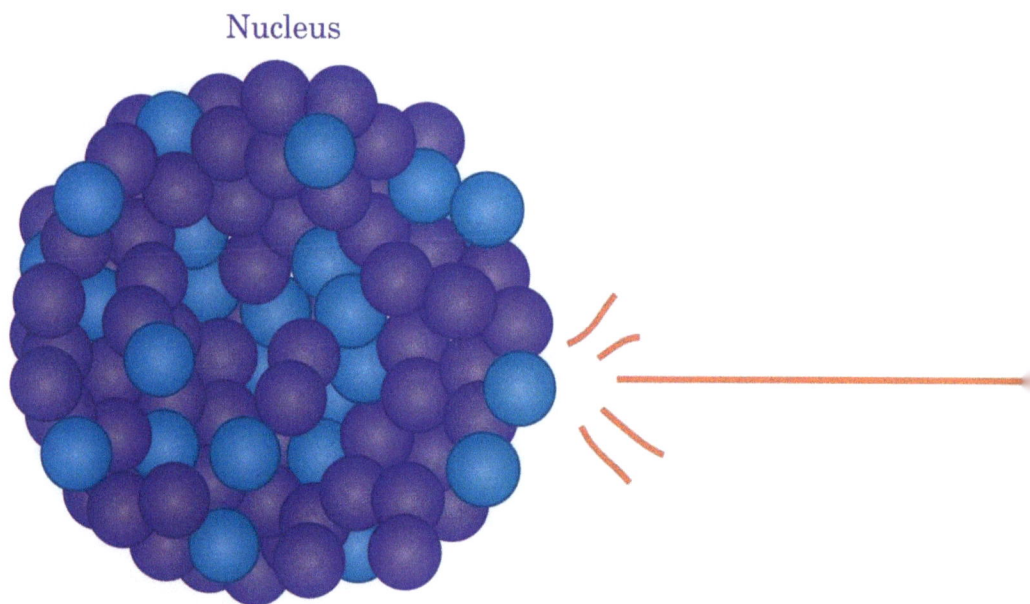

Alpha Decay. Even before Rutherford discovered the nucleus, he and his colleague Frederick Soddy at McGill University in Montreal, Canada, studied the chemical changes that accompanied radioactive decay. They discovered that when a radioactive element emits an alpha particle, it produces a "daughter" element that has an atomic number two lower than its parent and is lighter by four units in atomic weight. That told Rutherford that the alpha particle was a helium atom without its electrons (or, as they would later call it, a helium nucleus). This illustration shows an emitted alpha particle and the resulting newborn daughter nucleus, including the protons and neutrons that we now know are part of both.

Alpha particle
(2 protons, 2 neutrons)

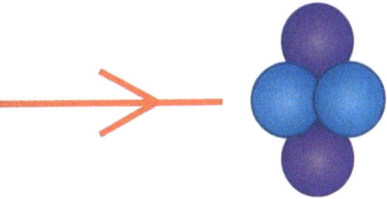

atomic mass. Radioactivity, however, comes from within the nucleus and transforms one atom into another. Unfortunately, for those people who hoped to achieve the alchemists' dream of transforming lead into gold, radioactive transformations went the other way, beginning with rare and valuable elements like uranium and ending— often billions of years later—with much less valuable lead.

Protons and Neutrons

Once Rutherford discovered the nucleus, he naturally drew on his work with radioactivity when thinking about what was inside nuclei. No one questioned that the hydrogen nucleus was a single proton, and Rutherford's studies had shown that alpha particles were helium nuclei, each of which carry two basic units of positive charge but have an atomic mass of four units. Looking at the atoms that produced the radioactivity, some scientists suggested that the extra mass in a nucleus was due to extra protons plus an equal number of electrons. Rutherford didn't agree, and in 1920, a year after replacing the retiring J. J. Thomson as leader of the Cavendish Laboratory at Cambridge University, he explained it this way. An electron inside a nucleus would experience a powerful electrical attraction to any proton it would encounter, and the pair would quickly bind together in a single, electrically neutral unit, a subatomic particle he called a neutron. He theorized that the alpha particles were made of four subatomic particles, two protons each with one unit of positive electric charge plus two uncharged neutrons.

Rutherford had already established that a radioactive atom that emits an alpha particle experiences a decrease of two in atomic number and four in atomic mass. Now he was saying that those decreases were due to a helium nucleus—a unit of two protons and two neutrons—bursting out of the larger unstable nucleus of a radioactive atom. He explained beta emission as the result of the splitting of a neutron into a proton and an electron. Since electrons are so light, the new nucleus would have approximately the same atomic mass, but its atomic number would increase by one. Later research showed that Rutherford was right about alpha emission and almost right about beta emission. Every neutron that decays into a proton and a beta particle also emits an even tinier particle, a nearly massless subatomic sprite called a neutrino.

Even scientists who agreed with Rutherford's theories didn't accept them without proof. Show us neutrons, they insisted, and Rutherford set out to do just that. It took more than a decade, during which physicists devised many new devices and techniques to observe the paths of subatomic particles through matter, though not the particles themselves. Those techniques were very effective for revealing the passage of electrically charged particles, but the neutron, if it existed, would remain invisible. Finally in 1932, James Chadwick (1891–1974), one of

Neutrino

Before

Neutron about to decay

After

Proton

Electron

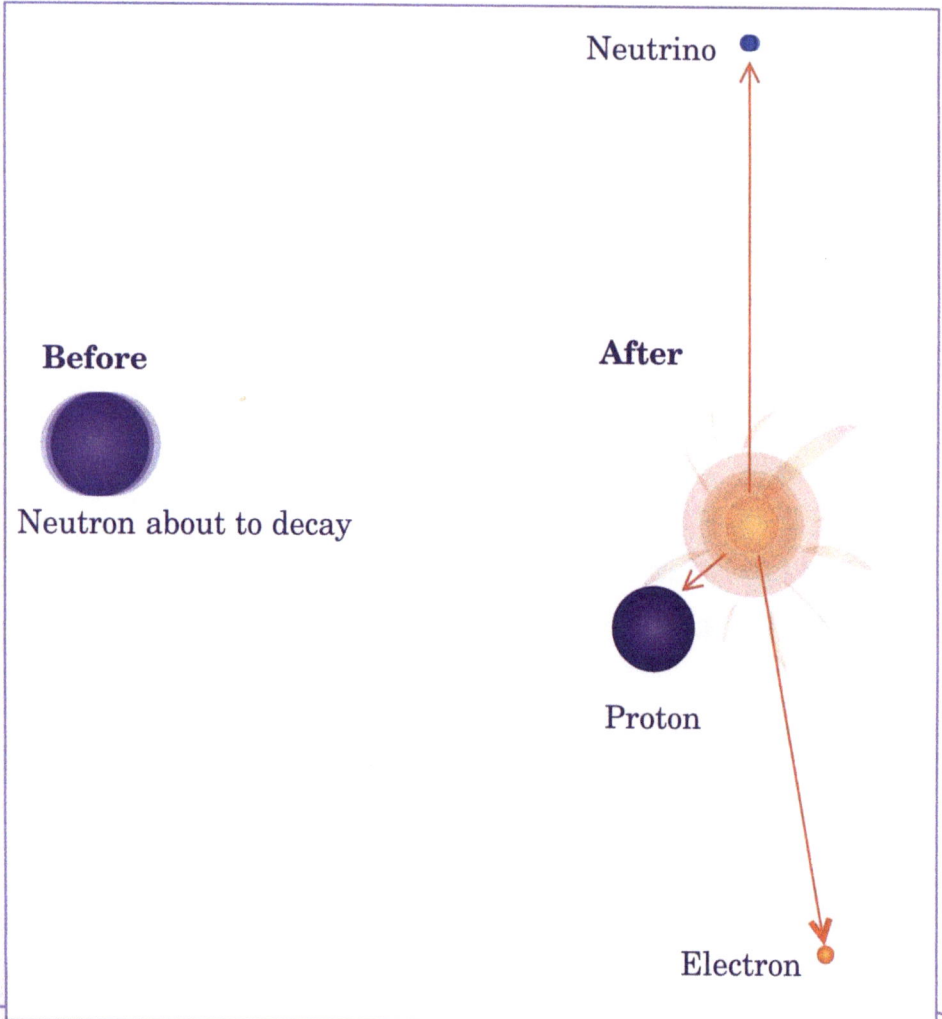

Beta Decay. At McGill, Rutherford and Soddy found that beta decay is accompanied by chemical change. As Rutherford came to understand more about nuclei, he realized that beta decay corresponded to the transformation of a neutron into a proton and an electron (the beta particle). The daughter element would thus have an additional proton, which matched the discovery that its atomic number increased by one over its parent. The electron was too light to make any difference in atomic weight. What no one yet knew was that, as shown here, a tiny neutral particle called a neutrino is always emitted at the same time as the beta particle.

Rutherford's colleagues at the Cavendish Laboratory, figured out a way to detect neutrons indirectly but convincingly. The basic structure and components of atoms as we now know them—positively charged nuclei of protons and neutrons carrying most of the mass while filling only about a ten thousandth of the atom's diameter, surrounded by light electrons—had been established.

Relativity, Quantum Mechanics, and Nuclear Forces

While Rutherford concentrated on the nucleus, two other revolutions in physics were also taking place: quantum mechanics and relativity. Together, they transformed our scientific understanding of matter and energy and space and time. Albert Einstein had a hand in both. Amazingly, he published his groundbreaking ideas in both fields in 1905, the same year that his explanation of Brownian motion proved that atoms and molecules are real.

The most famous scientific equation of all time is probably Einstein's $E = mc^2$, which he

Neutron Finder. Although Rutherford's prediction of the neutron was widely viewed as a sensible theory, scientists needed experimental evidence before they could accept it. Detecting a neutral subatomic particle was difficult, but James Chadwick, shown here, succeeded in doing so in 1932.

wrote almost as an afterthought to his theory of relativity. Just as relativity uncovered surprising relationships between space and time, this equation expressed the unexpected fact that mass (m) and energy (E) are two sides of the same coin. They are measured in different units, so we need a conversion factor to match them up, just as you might change a measurement in meters to inches by multiplying by 39.37. Nature's conversion factor between mass and energy is the speed of light (c) multiplied by itself (or squared). As scientists began to measure the masses of protons, neutrons, and the nuclei of different isotopes, they began to see the power of that simple equation. For example, when a radioactive nucleus emits an alpha particle, the mass of the daughter nucleus plus the mass of the alpha particle add up to less than the original mass. The missing mass turns out to be exactly the energy carried by the alpha particle.

Einstein's third great idea of 1905, and the one that led to his winning the Nobel Prize in 1921, changed the relationship between matter and energy in another way. He looked at two puzzling recent discoveries. The first was the photoelectric effect, in which light could knock electrons free from metals but only if its frequency was high enough. The second was an odd idea developed a few years earlier by Max Planck in his calculations of the spectrum (the mix of colors) produced by a hot body. Planck's equation depended on having light energy coming not in smooth waves like water, but in a stream of packets called quanta. He didn't believe that quanta were real, but they fixed a serious problem with his calculations. Einstein realized that the photoelectric effect was evidence that Planck's quanta—which he named photons—were real.

Other scientists soon recognized that all subatomic particles are quanta, just like photons, and they behave like waves at the atomic scale. They developed a new field called quantum mechanics that described the situation. For example, electrons in atoms had certain allowed wavelike states, each corresponding to a certain energy level specified by four "quantum numbers." Quantum mechanics worked spectacularly well for hydrogen, and when scientists imagined building larger atoms by adding more protons and electrons to them, they discovered something even better: The atoms had periodic properties. More than fifty years after Mendeleyev's dream, the periodic table made sense!

Quantum mechanics also changed the way science understood the basic forces of nature. For example, the laws of electromagnetism were recast in a new mathematical form called quantum electrodynamics, developed in the 1940s, which explains attraction and repulsion as the result of exchanging photons between electrically charged quanta, such as electrons and protons. In fact, in the quantum world, all forces are the result of such particle exchanges—and that takes us back into the nucleus. If protons repel one another by exchanging photons, what keeps the nucleus from blowing itself apart? There must be another force at work inside the nucleus, and it must have something to do with both protons and neutrons.

That force is known as the strong nuclear force, or simply the strong force. (There's another force called the weak nuclear force that explains beta emission.) The strong force must have some unusual properties. Unlike electromagnetism, which extends its influence a long way, the strong force must become insignificant beyond nuclear distances. Otherwise, nuclei of different atoms would be drawn together, and the universe would collapse into one giant atom. It must be much more powerful than the electromagnetic force within the nucleus, yet there must be a limit to its power outside the nucleus so that particles are not crushed to nothingness.

Such a force can be understood by mathematical approaches that are similar to those used in quantum electrodynamics, with a few differences. Nucleons—protons and neutrons—attract through a property that physicists call color, the strong force equivalent of electric charge, and instead of trading massless photons, they exchange particles called pi mesons that have a mass of about 250 times that of an electron.

How important and powerful is that strong force? It is responsible for all the elements and the energy of the stars. That's the subject of the last chapter of this book, where, at last, the proton begins to shine!

Chapter Four

A Universe of Protons

Living on a small rocky planet with some large puddles of salty water called oceans on its surface, it's hard to realize that, except for hydrogen, most of the atoms around us are uncommon. The best current scientific understanding of the universe is that it burst into being 13.7 billion years ago in an event commonly called the big bang. It began as seething, expanding matter and high-energy photons. In a short time, most of that matter became the subatomic particles we now know so well—protons and electrons in equal numbers plus a much smaller number of neutrons and alpha particles. None of the atoms with heavier nuclei familiar to us on Earth—carbon, nitrogen, oxygen, aluminum, iron, and so forth—were anywhere to be found.

Even today, most of the matter in the universe is hydrogen, and thus most of its mass is from protons. That is even true of our bodies, although they also contain plenty of neutrons. Most of the atoms in our bodies are hydrogen, but those atoms are combined with heavier atoms in compounds, such as with oxygen in water. Those heavier atoms contribute most of our bodies' masses. Our bodies also need energy extracted from plant and animal matter. That energy originally came to Earth from sunlight. So what do those facts have to do with protons? Everything! The sun is a star, and besides making light, stars are the factories where nature has cooked its heavier nuclei. The cooking process is called nuclear fusion, and it begins with the most common nuclear raw material available in the universe—protons!

Furnaces of Creation

During the big bang, much of what happened was the opposite of what goes on during radioactive decay. For example, during beta decay, a nucleus emits a beta particle (an electron) and a neutrino when one of its neutrons gives up some of its mass to become a proton. To reverse the process and make a neutron from a proton, an electron, and a neutrino, those three particles would need to come together with at least as much energy as was released in the decay process. During the early stages of the big bang, protons, electrons, and neutrinos appeared first, and everything was close together and moving fast enough to make some neutrons. But the expansion was so rapid that neutron making ended quickly, leaving most matter in the form of protons, electrons, and neutrinos.

Besides making neutrons, the early universe was also hot and dense enough to cook up a few other nuclei. For instance, some protons and neutrons paired off to form the nucleus of a heavier isotope of hydrogen with mass number 2 called deuterium (denoted as D or H^2). Two isotopes of helium also formed. The more common helium nucleus was the alpha particle (two protons and two neutrons or He^4), but helium nuclei with only one neutron (He^3) were also stable. Very small numbers of nuclei of other light elements also formed in the big bang, but not enough to be concerned about. When the first burst of matter-creation ended, the nuclei picked up electrons. At that point, most matter in the

Deuterium
nucleus

Tritium
nucleus

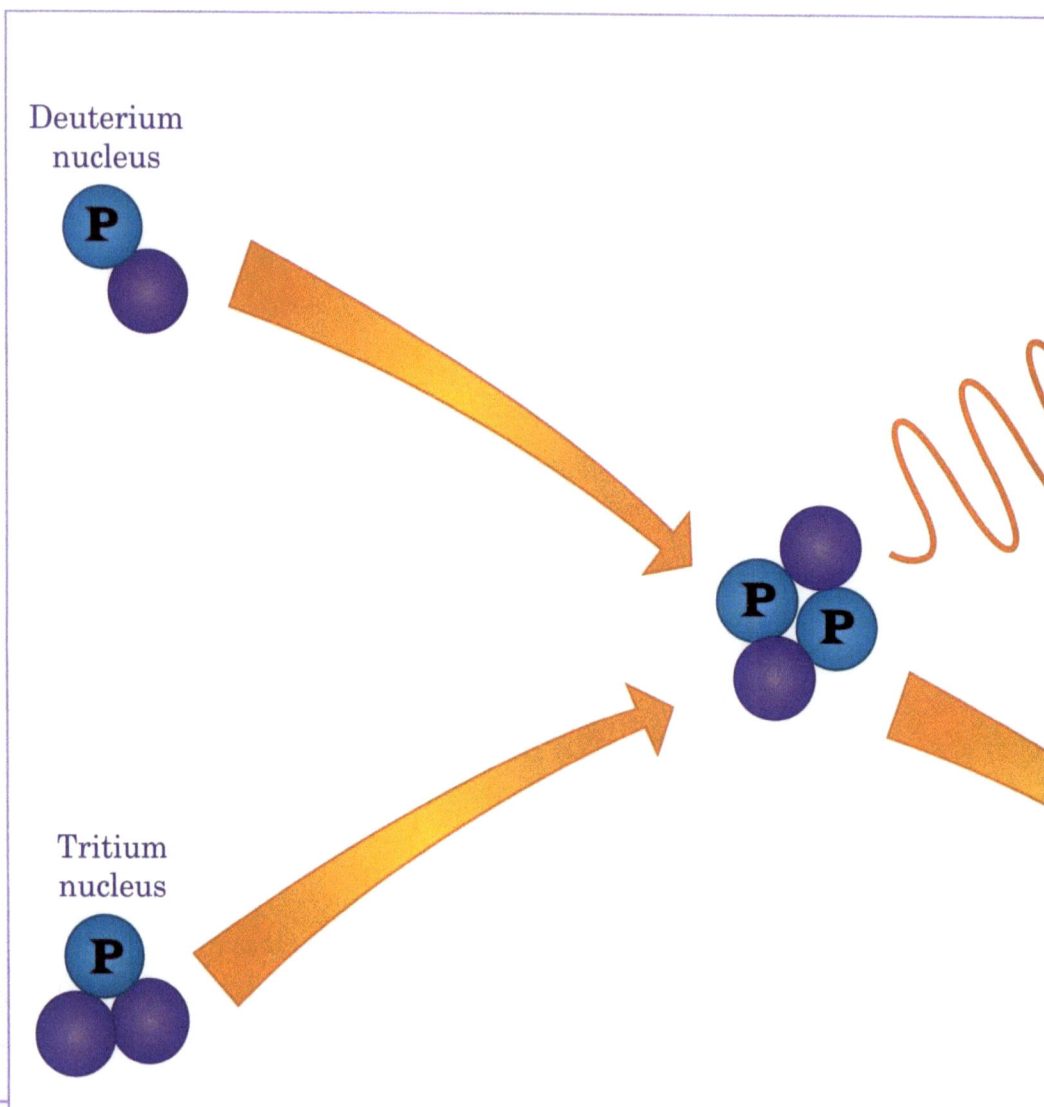

Stellar Energy. A star produces its energy from a process called nuclear fusion. Many different fusion reactions take place, but this diagram shows the one commonly known as D-T fusion. A deuterium nucleus (a heavy isotope of hydrogen with one proton and one neutron) fuses with a tritium nucleus (an even heavier isotope of hydrogen with one proton and two neutrons), producing an alpha particle (a helium-4 nucleus), a free neutron, and a photon of light energy. Later in the lives of stars, other fusion reactions take place that create all the heavier nuclei in the universe.

Photon of
light energy

Free
neutron

universe consisted of hydrogen atoms, and almost all the rest consisted of helium atoms.

Once atoms formed, gravity could prevail over the more powerful electric force. Bare nuclei would repel each other electrically, but not neutral atoms. In some regions of the early universe, by pure chance, more atoms than average had formed. There, gravitational attraction began to draw the atoms together, slowly at first and then more rapidly. Eventually, large numbers of atoms were coming together at high speed, sometimes heating up enough to strip away their electrons. Because of their high speed, not even electrical repulsion could stop the nuclei from getting close enough so that nuclear forces would take over. If circumstances were right, the nuclei would join together, or fuse.

Unlike the process that created neutrons from protons, electrons, and neutrinos in the early instances of the big bang, nuclear fusion events can sometimes release energy instead of absorbing it. That occurs when they make a particularly stable nucleus. For example, two deuterium nuclei (or deuterons) release energy when they fuse to form an alpha particle. Before and after the fusion event, there are two protons and two neutrons, but the alpha particle has less mass than the two deuterons. The difference in mass is released as energy. Fusion events like that are what power stars like the sun, although the precise fusion reactions that take place are a bit more complicated than this example. Still, if gravity pulls enough hydrogen together in one place, fusion begins and continues until nearly all the protons have fused into alpha particles. While the star is "burning" its proton fuel, the outward pressure due to the heat counteracts the gravitational force that has been collapsing the matter. When that fuel is finally used up, the collapse begins again, which raises the temperature of the star. Suddenly, there is enough energy to ignite other fusion events, such as the combination of three alpha particles into a carbon nucleus (C^{12}), which may fuse with another alpha particle to make an oxygen nucleus (O^{16}) at a still higher temperature. Those processes also release energy, but each step requires a higher temperature to ignite it. Depending on its size, each star stops all fusion at a different point and becomes a slowly cooling cinder. For example, in about 5 or 6 billion years, our sun will end up as a ball of hot carbon.

Larger stars follow other paths, but once a star is done fusing its nuclei into iron, additional fusion requires more energy than it produces. Iron's atomic number is only 26, so where did Earth get all its elements with much larger numbers of protons? (Uranium, the naturally occurring element with the most protons, has ninety-two of them). The formation of those elements requires a powerful event known as a supernova explosion, which occurs only for the largest stars. Those stars have so much mass that gravity squeezes their nuclei together until they explode with great violence. The event produces so much energy that it forces nuclei to fuse that would not ordinarily do so. You might think that the products of fusion would all be unstable and blow apart as soon as the pressure is released. Many of them are, but some are like water in a deep well on top of a mountain. Once the water is in the well, you need a lot of energy to pump it out so it can run down to the valley. A few nuclei are radioactive—unstable but lasting a long time. They are like water in a shallower mountaintop well where severe storms come by from time to time. If you wait long enough, a storm is sure to strike and blow the water out.

Protons in Research and Technology

Understanding the nuclear processes in stars took a lot of research, and people involved in research often think about ways they can apply their discoveries. For instance, fusion energy has already been used in fearsome military weapons known as hydrogen bombs, or

What a Blast! Scientists have been able to create weapons called hydrogen bombs, or thermonuclear devices, that create huge destructive blasts of fusion energy. These bombs, which are vastly more powerful than the "atomic bombs" dropped on Japan in 1945 to end World War II, have been tested (as shown here) but never used as weapons.

thermonuclear devices. Might there be a way to make an electric power plant operating on the same principle? It would be like capturing the energy of the stars in a bottle. In fact, engineers have been trying to build fusion reactors for electric power for almost fifty years, but the biggest problem is finding a bottle to contain

Hot Doughnuts! This photograph shows the inside of a doughnut-shaped controlled-fusion device called a tokamak, designed in the hope of creating inexpensive and environmentally friendly electricity. Powerful electromagnets squeeze a beam of fast-moving nuclei together, and some of them fuse to produce energy. Unfortunately, it is very difficult to keep the million-degree beam together long enough to produce more energy than the device consumes, but engineers are still seeking ways to make fusion power a reality.

the super-hot gases. Since the fusing nuclei are electrically charged, they respond to magnetic fields, but even a slight instability is enough to disrupt the process before enough fusion energy is released. It takes electrical energy to make the powerful magnetic fields needed, and so far the most successful devices have only succeeded in producing about as much energy as it takes to run the reactor.

Another fusion power technique is called inertial confinement, and it avoids the need to create a powerful magnetic field. Instead, it blasts a pellet of hydrogen-rich material on all sides with a laser or similarly intense energy source, hoping to create a situation similar to the process by which material is drawn inward by a star's gravity. The object is to quickly create enough fusion reactions before the heat blows the material outward again. It seems like a good idea, but so far it has been very difficult to achieve in practice.

As scientists have investigated subatomic particles and new forces, such as the strong and weak nuclear forces, they have created machines that make beams of protons. These beams have a number of uses, including the treatment of some forms of cancer and the detection of very small amounts of certain elements in a sample of material. But probably the most interesting use is what happens when two proton beams are accelerated to nearly the speed of light and then allowed to collide. The result is a shower of new and very unusual particles that reveal inner secrets of matter that might have surprised Ernest Rutherford even more than the discovery of the nucleus did.

The scientists who create such powerful beams and machines have known for some time that there are even smaller particles within protons and neutrons called quarks, which stick together so tightly that they can never be detected separately. There's a lot you can learn about quarks, but that's another story for a different book in this series.

Glossary

alchemy A field of study that preceded chemistry through which many people hoped to transform less valuable metals into gold but never succeeded.

alpha particle or alpha ray A helium nucleus that is emitted from some radioactive elements.

atom The smallest bit of matter than can be identified as a certain chemical element.

atomic mass or atomic weight The mass of a particular atom compared to a standard that sets the mass of a C^{12} atom to be exactly 12. For a particular isotope, that value is approximately the number of protons plus the number of neutrons in its nucleus. For a naturally occurring element, that value is approximately the number of protons plus the average number of neutrons in the nuclei of naturally occurring isotopes.

atomic number The number of protons in the nucleus of an atom, which determines its chemical identity as an element.

beta particle or beta ray An electron that is emitted from some radioactive elements.

Brownian motion First observed by Robert Brown as the jiggling motion of a particle of dust or pollen suspended in a fluid; eventually shown by Albert Einstein to demonstrate the existence of atoms and molecules.

color A property of nuclear particles that makes them stick together, just as electric charge is a property of particles that makes them respond to electromagnetic forces.

compound A substance made of only one kind of molecule that consists of more than one kind of atom. For example, water is made of molecules that contain two atoms of hydrogen and one atom of oxygen.

electromagnetism A fundamental force of nature, or property of matter and energy, that includes electricity, magnetism, and electromagnetic waves such as light.

electron A very light subatomic particle (the first to be discovered) that carries a negative charge and is responsible for the chemical properties of matter.

element A substance made of only one kind of atom.

emission Sending out something that has been produced, such as sending out an alpha, beta, or gamma ray from a radioactive atom.

fusion The joining of two nuclei to form a new nucleus.

gamma ray A high-energy photon that is emitted from some radioactive elements.

molecule The smallest bit of matter that can be identified as a certain chemical compound.

neutron A subatomic particle with a neutral electric charge found in the nucleus of atoms.

nucleus The very tiny, positively charged central part of an atom that carries most of its mass.

periodic table of elements An arrangement of the elements in rows and columns by increasing atomic number, first proposed by Dmitry Mendeleyev, in which elements in the same column have similar chemical properties.

photoelectric effect A phenomenon in which light can, under some circumstances, knock electrons out of atoms. Einstein's explanation of this effect led to scientific acceptance of the photon as a particle and eventually to quantum mechanics.

photon A particle that carries electromagnetic energy, such as light energy.

proton A subatomic particle with a positive electric charge found in the nucleus of an atom.

quantum electrodynamics A formulation of electrodynamics that accounts for quantum mechanical phenomena and relationships, such as the dual wave-particle nature of matter and energy.

quantum mechanics A field of physics developed to describe the relationships between matter and energy that account for the dual wave-particle nature of both.

quark A sub-subatomic particle that exists in several forms that combine to make protons, neutrons, and some other subatomic particles.

radioactivity A property of unstable atoms that causes them to emit alpha, beta, or gamma rays.

scattering An experimental technique used to detect the shape or properties of an unseen object by observing how other objects deflect from it.

spectrum (pl. spectra) The mixture of colors contained within a beam of light, or the band produced when those colors are spread out by a prism or other device that separates the colors from each other.

strong nuclear force, or strong force A fundamental force of nature that acts to hold together the protons and neutrons in a nucleus.

theory of relativity A theory developed by Albert Einstein that deals with the relationship between space and time. Its most famous equation ($E = mc^2$) describes the relationship between mass and energy.

transmutation The transformation of one element to another by a change in its nucleus, such as by alpha or beta emission.

weak nuclear force, or weak force A fundamental force of nature that is responsible for beta decay of a radioactive nucleus.

For More Information

Organization

Lederman Science Center
Fermilab MS 777
Box 500
Batavia, IL 60510
Web site: http://www-ed.fnal.gov/ed_lsc.html
This museum is an outstanding place to
discover the science and history of subatomic
particles. It is located at the Fermi National
Accelerator Laboratory (Fermilab) outside
of Chicago.

Magazines

American Scientist
P.O. Box 13975
Research Triangle Park, NC 27709-3975
Web site: http://www.americanscientist.org

New Scientist (U.S. offices of the British
    magazine)
275 Washington Street, Suite 290
Newton, MA 02458
Web site: http://www.newscientist.com

Science News
1719 N Street NW
Washington, DC 20036
Web site: http://www.sciencenews.org

Scientific American
415 Madison Avenue
New York, NY 10017
Web site: http://www.sciam.com

Web Sites

Due to the changing nature of Internet links, the Rosen Publishing Group, Inc., has developed an online list of Web sites related to the subject of this book. This site is updated regularly. Please use this link to access the list:

http://www.rosenlinks.com/lsap/prot

For Further Reading

Close, Frank, Michael Marten, and Christine
Sutton. *The Particle Odyssey: A Journey to
the Heart of Matter.* New York: Oxford
University Press, 2002.

Cooper, Christopher. *Matter* (Eyewitness Books).
New York: Dorling Kindersley, Inc., 2000.

Henderson, Harry, and Lisa Yount. *The
Scientific Revolution.* San Diego: Lucent
Books, 1996.

Narins, Brigham, ed. *Notable Scientists from
1900 to the Present.* Farmington Hills, MI:
The Gale Group, 2001.

Strathern, Paul. *Mendeleyev's Dream: The
Quest for the Elements.* New York: Berkley
Publishing Group, 2002.

Bibliography

Cropper, William H. *Great Physicists: The Life and Times of Leading Physicists from Galileo to Hawking*. New York: Oxford University Press, 2001.

Nobel Foundation. *Nobel Lectures in Physics, 1901–1921*. River Edge, NJ: World Scientific Publishing Company, 1998.

Strathern, Paul. *Mendeleyev's Dream: The Quest for the Elements*. New York: Berkley Publishing Group, 2002.

Young, Hugh D., and Roger A. Freedman. *University Physics: Extended Version with Modern Physics*. Reading, MA: Addison-Wesley Publishing Co., 2000.

Index

About the Author

Award-winning children's author Fred Bortz spent the first twenty-five years of his working career as a physicist, gaining experience in fields as varied as nuclear reactor design, automobile engine control systems, and science education. He earned his Ph.D. at Carnegie-Mellon University, where he also worked in several research groups from 1979 through 1994. He has been a full-time writer since 1996.

Photo Credits

Cover, pp. 1, 3, 12, 14–15, 19, 20–21, 30–31, 34, 44–45 by Thomas Forget; p. 7 © Archivo Iconografico, S.A./Corbis; p. 9 © Jean-Loup Charmet/Science Photo Library; p. 10 © Sheila Terry/Science Photo Library; p. 25 © Peter Fowler/Science Photo Library; p. 36 © A. Barrington Brown/Science Photo Library; p. 49 © Corbis; p. 50 © EFDA–JET/Science Photo Library.

Designer: Thomas Forget; Editor: Jake Goldberg

www.ingramcontent.com/pod-product-compliance
Lightning Source LLC
Chambersburg PA
CBHW041723210326
41598CB00007B/764